Aquaponic Gardening For Beginners:

Raising Fish and Growing Vegetables in Aquaponics Garden

By

Erin Morrow

Table of Contents

Aquaponic Gardening For Beginners:

Raising Fish and Growing Vegetables in Aquaponics Garden

By Erin Morrow

Introduction

Aquaponic gardening is a symbiotic gardening system that relies upon both flora and fauna to create a balanced ecosystem for harvesting both fish and plant material. It requires a larger amount of setup and investment than a standard hydroponic system, but contains benefits over it that we will address later on, it as well is less dependent upon outside addition of nutrients or material to maintain its optimal operation level.

The key thing first when setting up an aquaponic system is determining what exactly is the goal for the system, are you looking to harvest a particular type of plant material and fish, and what are the requirements for those particular items. In this book we will focus on the use of tilapia as a fish stock and lettuce as the plant material source to simplify the process, but know that adding other vegetation or fish stock is not a complicated step once you have the system up and running, you just need to make sure that the nutrients for the plant are available and that the feed for the additional fish stock is present, otherwise it is a simple thing to add to the system once it is begun.

Chapter 1. Initial Requirements

There are a few requirements that are constant across all variations of the aquaponic systems, you will need a rearing tank for the fish to be contained within, a settling basin to catch loosened biofilm, settled particulates and uneaten food, a bio filter for the bacteria to process the ammonia into nitrates for use by the plants, a hydroponic subsystem where the plants themselves will be residing, and finally a sump for pumping the water back to the beginning of the system to continue to transition the water chain.

Now as I said from the beginning we will focus on tilapia as our fish stock so we will need to make sure we have a proper sized tank for that fish stock. For a standard tank you are looking at 1 gallon (3.8 liters) of water per 0.5 lbs. (0.23 kg) to 1 lbs. (0.45 kg) of fish with proper filtration and aeration of the tank. For your tilapia we will work with an Oreochromis mossambicus, or what is commonly called a Mozambique Tilapia, it weighs around 2 lbs. (0.9 kg) as an adult, has high tolerance for stocking and is very tolerant of a range of conditions for habitation, making it a perfect fish for an aquaponics gardening system. Now with its weight we will need to calculate an accurate amount of water, we will run with the calculation of 1 gallon per 0.75 lbs. of fish for a safer estimate. You will want to have between four to six fish at the start up point and so you will need sixteen to twenty-four gallons of water for your rearing tank startup, you can use a common aquarium for this or you can custom make the container if you wish. Custom made containers are easily constructed from a large plastic barrel sliced in half, with one half being for the rearing tank and the other can be used for the hydroponic subsystem if you wish.

Once you have the rearing tank set up, you will need to make sure the pH levels are good and that it cycles with the rest of the system so we still are not yet to the point of actually adding in the fish, just getting the rearing tank itself set up. Once you have a tank of proper size you will need to make sure you have a filtration system in place to promote bacteria growth and remove the waste and ammonia from the rearing tank itself. An under gravel filter is easily obtained for an aquarium base and will meet the need we have at getting the material out that we need out, and promoting bacterial growth which convert the ammonia into a nitrate for use by the plants.

You will also want to obtain an air pump with air rock to place within the tank to ensure proper aeration of the tank for the fish stock being placed there. As a helpful hint on this part and with all tubing being used, make sure it is an opaque tube so to not allow light in, if there is light getting in then you will have algae growth within the tube which will cause clogs in the system and lead to failure, or a large amount of work cleaning or replacing the tubes themselves, so ensure the opaque tubes to save yourself many future headaches.

Once the rearing tank is fully setup with filtration system and standard air rocks we move on to the plants and their grow bed. Now there are numerous ways to do this, you can use a gravel or clay base where they are flooded with the water, or you can do a floating medium that does not require the gravel but does require another tank and a little more hoses. For a gavel base you will need a container that can hold both the gravel / clay and the plant itself along with a pump to move the water into the container and allow it to drain out, the less intensive and more productive measure though is the floating bed, which is what we will address.

You will need another tank, an aquarium the same size as the one you use for the rearing tank would be ideal, giving a deep level for the plants to expand their roots. To begin the deep water culture setup we are going to need to set up the tank similar as the rearing tank, you will add an air stone with pump to keep the tank aerated properly for the plants, and you will need to obtain Hydroton, which is easily available at any hydroponic type store or online ordering if unable to locate in a nearby store. Once you have your Hydroton you will need to wash it a couple times to remove any excess clay dust still on it to prevent contamination of your tank. Next you will need to get an organic plug, which is just a planting medium for the seed to be placed within to grow, it comes loaded with nutrients to help start your plant along as well, and you place all of this within. Then you will need a net pot, which is basically a pot but with hole throughout the bottom to allow the roots to expand out through the sides and access more nutrients.

You will need to have six of the net pots for this starter deep water culture, you will fill each a quarter of the way with the Hydroton then place the organic plug in the center and surround it with the Hydroton to secure it in place, if there is no pre-drilled hole in the organic plug then you will need to remove one fourth an inch deep in the center to plant the seed in. Once that is all completed your plants are ready for being placed in the system. You can use a Styrofoam board to float the pots, just get a board slightly smaller than your tank measurements, then cut holes in it to set the pots within, make sure you do not pinch your tubing for the aerator as well, and it should float the pots at the surface easily to reduce trying to find other means to suspend them.

Now before you add the plants you need to make sure you have the pump system setup to return the water back to the rearing stock itself, remember this is a cyclical system, one pump moving the nutrient rich water from the rearing tank to the plants, and then another pump moving that water back to the rearing tank. The easiest way to do this is to set both tanks side by side with the pumps situated at each end transferring the water flow. You will also want to make sure to use opaque hoses for all the pumping, it will prevent algae growth and save you many headaches in the future.

Once you have both pumps setup with the water flowing between both tanks at a sustained rate, let it run for a couple days to make sure that there is no serious faults or problems that will need to be addressed. If everything runs smoothly with no problems for a couple days with no fish or plants introduced then it is time to introduce the fish. Add them in at a young age, you will start with fry not adult fish, since you are starting with seeds and not transplanting you will need to keep the system on a balance. Add in the fish and allow the system to run for another two days, checking the pH level and making sure it stays in the 6.8-7.0 range.

After that period you can introduce your plants, make sure the Styrofoam float does not pinch the hoses and that water flow is still good, and allow this to go for a couple days and again check the pH levels to ensure they are not changing. If they are you can obtain safe chemicals to adjust the pH levels from an aquarium store or hydroponic store that will not affect your fish or plants. Once you have stabilized the pH levels and the system has been running for a week you should be able to relax and only check the pH levels weekly instead of every couple days. Once a month you will need to make sure the water levels are good, if they are low you will need to add in de-chlorinated water, it is very important to remove the chlorine from your water, since it will not just affect the fish and plants but also kill the bacteria that are vital the entire function of the system, the ones converting the ammonia from the fish into nitrates for the plants. Along with tending the water level you will want to use a siphon cleaner to remove any excess waste in the rearing tank that has not been removed, you can also add in a plecostomus, a type of fish that is sometimes referred to as a janitor fish since it cleans waste by products out of the tank from the other fish, that will help clean the rearing tank as well, further reducing maintenance on your part.

Chapter 2. Feed For The Fish

Now the main source of work for you within an aquaponic system is the feed material for the fish stock being used. Constantly buying feed for the fish can be problematic for some, and you need to make sure you are getting the proper type of feed as well to ensure healthy growth in the fish of choice. In this example we are using the tilapia which has a diet that is omnivorous in the wild but within the fish stock we will want to use a feed pellet that can be obtained from an aquarium store or aquaponics store, or if you have the space you can set up a compost for growing black fly larva which are good feed sources, but you would also need to grow some duckweed in the hydroponic section to transplant to the rearing tank for them to feed on as well.

With each species of fish being used you will have to make sure to find out its nutritional needs so that way you can either purchase the proper feed source for them, or find a means to grow it in your local environment for feeding the fish stock as needed. This is the primary source of expense and effort when the system in is full function, since the feeding will vary based upon the size of the fish stock, as well as the species being used, it can vary from a short task to a long chore depending upon how many fish you have and is the reason you want to make sure you have easy access to the rearing tank during set up, so that you can feed them without even more additional work on you.

Chapter 3. Benefits of an Aquaponic Garden

Now that you have your system fully set up and running, you will get to see the benefits of such a system. Standard hydroponic systems are requiring a constant supply of nutrients added to the water to maintain the plant material growth, while an aquaculture has to continually remove water to avoid toxicity of the water for the fish stock, but with an aquaponic system you have neither drawback thanks to the nutrient rich water from the aquaculture tank being used to feed the hydroponic tank, you remove the most expensive and work intensive aspects from both of those techniques while gaining the benefit of having both systems running for you.

We listed one method for setting up a system that will give you a nice smaller garden that will produce fish and vegetables for your consumption, but that is not the only method, there are numerous ways with all having the same benefits, no weeding, no fertilizers and no back pain inducing bending over all day. One such method would be the recirculating hydroponic system, where you grow your plants in a gravel medium that is flooded with the water from the rearing tank and then drains down back to the rearing tank from above, it reduces the pumps needed for the process but does not have the same yield on average that the deep water culture method has, none the less, you use one tenth the amount of water to grow the plants than if you were to use standard soil based gardening, and you can actually grow the plants in more dense packing since the roots will not have to support the plants weight they will be able to grow narrow and deep into the deep water culture and let you stack the plants closer together gaining more harvest for less space. That benefit also applies to the fish rearing tank, since in an aquaponic system the stocking density can be quite high for the fish and still maintain a healthy growth rate for them. The benefits from the stocking density are truly impressive, in the wild you would not be able to maintain anywhere near the same number of fish within the space being used, the food supply would not be sufficient and the waste produced would make the water toxic for them in short order, but thanks to the aquaponics system cycling that ammonia out and bringing back in fresh water you will have numerous healthy fish with minimal effort.

Chapter 4. Pest Control

Even with a highly efficient and well regulated system there will be pests, such as caterpillars which can be controlled with a simple spraying of Bacillus thuringiensis, a soil borne bacteria that has insecticide properties and will help prevent the caterpillars from showing up. You may also encounter sap sucking insects which can be dealt with using a chili and garlic spray, make sure to not use excessively, they will affect the water if over used. Now if you happen to have slugs and do not want them, then you can use a small tray filled with beer, it will attract them and they will become trapped and drown. As well you may have to address molds or fungus growths, which are best addressed with a potassium bicarbonate spray, which may be useful to add to the tank as well to help the pH levels stay elevated as well as adding needed nutrients, again do in moderation to not affect the system excessively.

You would also have the normal issues with the fish themselves, making sure to check them for things such as sick / white spot, fin rot and such. The treatment for these common conditions found in many aquarium based fish are dealt with the exact same, since there are minimal drugs that actually can be used for the treatment of fish, or the diseases within them, you will want to make sure if you notice any of the common diseases afflicting a fish to remove that fish and place it in an isolation tank for treatment and to monitor the other fish more closely to ensure the disease does not spread.

Chapter 5. Expanding The System

Once you have your basic rudimentary system up and running you may at one point wish to expand it, by adding new species of fish or vegetation, which is not a bad idea but must be thought through fully before you do so. If you wish to maintain the same size tanks but to varying up the fish species then you need to make sure you are using a species that has the same climate needs as the current species as well as similar sizes so to not exceed your rearing tanks capacity. You will also want to make sure to have the new species kept in an isolation tank for a few days before introducing it to the system to make sure there is no indications of diseases that would be brought into the system from the new fish species.

As well if you are introducing adult species you will want to make sure the temperament of the species will fit within the environment, you do not want your fish attacking one another. Now if you are looking to introduce new plant species you will also have to look through the details, how large will the plant be, will it need the same nutrient resources or will you have to add in additional items for that particular species. It sounds far more complicated than it actually is, but you will want to make sure you address all of these before expanding upon your system since you will not want it to go to waste from a simple mistake.

Chapter 6. How Aquaponics Works

The system itself is a nearly self-sustaining system, the fish consume food you add in, which then they produce waste, that waste is filled with minerals and nutrients and ammonia. The ammonia itself can become quite toxic to the fish if not addressed which is what the bacteria are for, they convert the ammonia in to nitrates, well the Nitrosomonas convert the ammonia in to a nitrites which is then converted in to a nitrate by Nitrobacter, which is then used by the plants for growth.

The process is called the nitrogen cycle, which allows you to use the fish waste product to produce a fertilizer for the plants using bacteria that feed off the waste itself. Once the ammonia has been converted and the plants have absorbed what nutrients they can, the water is then transported back to the rearing tank which the fish add more ammonia to it and the cycle continues, needing you to only add water to combat evaporation and transpiration by the plants.

Chapter 7. Types of Fish That Can Be Grown in Aquaponic System

Within an aquaponic system you are able to rear numerous fish species, the above guide was in reference to a tilapia rearing stock which are the more tolerant species only requiring you to have a warm temperature for them, which is easily obtained if you are doing it indoors. Now that is not to say you are limited to only tilapia, the options are quite varied, such as goldfish or koi if you are not looking to actually eat the fish and just want to maintain a system, if you are wanting a system but do not plan to regulate the water temp or are unable to do so reliably then you would use a bluegill or catfish, since they have a much higher tolerance in temperature range.

Now you can also raise some things such as rainbow trout with these aquaponic systems if you maintain a colder climate for them, if you are in a warmer region you can use a barramundi, if you happen to have a temperate climate that varies with the seasons then you can actually alternate the species being reared through the year with the changes in climate, allowing you to have a variety of fish throughout the year for consumption.

Chapter 8. Types of Plants That Can Be Grown in Aquaponic System

Our above tutorial used lettuce as an example plant, just for simplicity of the guide, but that is not what you are limited to when growing vegetation with an aquaponic system. With experience and creativity there is no real limit to what you can grow in an aquaponics system when it comes to plants. With a large enough system you can grow dwarf fruit trees even to harvest oranges or grow cauliflower along with cabbage, lettuce (as mentioned above) and some beetroot, all in the same aquaponic system.

The key thing to consider is what plants you want to eat, and then to make sure you have a large enough selection that you will not harvest them all at the same time, you will need to make sure to have plants growing at all times in your aquaponics system, otherwise there will be nothing to remove the nitrates produced by the bacteria which will affect your fish, and could lead to ruining the entire system.

The only requirement for the plants if that you meet the temperature for their growth requirements and proper medium usage. You would have to research what type of medium would work best for each type of plant if you are going for the more exotic or larger plants. The Hydroton as mentioned above works wonderfully for the smaller scale plants, but you may need something in addition to that for the larger plants, it does depend upon the plant itself. Now as for a list of the plants that are commonly used in an aquaponic system you are looking at lettuce (as mentioned above), cabbage, bell peppers, tomatoes, okra, and most any leafy green vegetable you can think of. So you really are not limited in what you can grow in this medium, just figure out which ones you would like to have available to you, then you can setup a system that will grow it for you.

Final Words

Once your system is fully up and running and have been in use for six months you will honestly wonder why you never tried this time of gardening before, it has all the benefits of hydroponic and aquaculture without the messing drawbacks, as well as giving large harvest yields with much less work, as well as having a lower space requirement for growing the plant material as well. No longer will you need to have a huge space to till, and weed and fertilize, now you need only some large plastic containers or aquarium tanks that you can convert into a lush and vibrant garden full of delicious plant material and fish stock tanks full of delectable tasting fish.

This technique is such a successful method for farming both fish and plant material it is becoming a main stay in industrial level farming as well, since the conservation of resources used to raise both fish and plant are far below that of the standard process it will be far more profitable for them to use this safer, more natural ecosystem style farming method.

This aquaponics system can ever be used for just basic decoration as well, using plants such as roses or ivy for a nice appearance or smell, and a non-edible fish species such as koi in the aquaculture tank, you will be able to have a nice symbiotic ecosystem set up within your own house or garage to not just enjoy the smell and appearance but also to educate your kids with. So don't wait any longer, follow this guide and you can have your own aquaponics system setup and running in no time, with either fresh plant material and fish or neither ready for you in a few short months and will save you money at the grocery from that point forward.

I want to personally thank you for reading my book. I hope you found information in this book useful and I would be very grateful if you could leave your honest review about this book. I certainly want to thank you in advance for doing this.

If you have the time, you can check my other books too.

.

CPSIA information can be obtained
at www.ICGtesting.com
Printed in the USA
LVHW042253100119
603542LV00020B/1397/P

9 781681 270043